Country Crafts in Pictures

J. E. Manners

David & Charles Newton Abbot London
North Pomfret [VT] Vancouver

By the same author
Country Crafts Today

ISBN 0 7153 71479

Library of Congress Catalog Card Number 75–43205

© J.E.Manners 1976

Photoset in Ten on Eleven point Bembo
and printed in Great Britain by
Redwood Burn Limited, Trowbridge & Esher
for David & Charles (Publishers) Limited
Brunel House Newton Abbot Devon

Published in the United States of America
by David & Charles Inc
North Pomfret Vermont 05053 USA

Published in Canada
by Douglas David & Charles Limited
1875 Welch Street North Vancouver BC

Contents

Introduction

Up till the end of the eighteenth century, most villages in England were self sufficient both in food and in the provision of the everyday goods and services required by an agricultural community. These latter needs were met by local craftsmen. However, with the advent of the Industrial Revolution and the consequent improvement in communications, the drift from the countryside began. Mass-produced factory goods became available everywhere and rural craftsmen found themselves no longer indispensable.

Since those days, numbers have dwindled drastically, few crafts surviving unscathed. The long apprenticeships needed to learn the craft skills do not attract young men to the trades when a quicker and easier living can be earned working a five-day week in a factory. While a few like the independence of working for themselves, many more are put off by the insecurity. No money comes in if the craftsman is sick or on holiday; woodland craftsmen particularly live a lonely life and conditions are less than pleasant out in all weathers. Labour costs have escalated to such a degree that craftsmen are afraid of pricing themselves out of the market, and the older men who remember the depression of the 1930s are most conscious of this. With little new blood attracted into the craft industries, the average age of existing craftsmen is high, and there is a danger that craft skills, passed on from generation to generation, may well die out.

There are some bright spots, however, as the affluent society has brought relative prosperity and security to such crafts as saddlery, farriery and thatching. More leisure time means an increased demand for sports equipment from cricket bats to harness, and leisure pursuits such as spinning and weaving are being practised on an increasing scale. Basketry, hurdle-making and coopering are declining but those left plying these trades are doing well. Similarly, wrought-iron workers, hand brick-makers and dry-stone wallers are being kept busy, especially when they have established a reputation for their skill.

Almost as a reaction against the instant, assembly-line goods of the 1970s, there has been a revival of interest in craft work and a new appreciation of individual skills. This book attempts to bring a little of the limelight to craftsmen who ply their trades unobtrusively, illustrating them at their work and giving some hint of the enormous skill and care which goes into the making of their products.

Trees and wood

Wood plays a dominant part in the work of a large number of rural craftsmen. The hardwood of our native deciduous trees is the preferred material although softwood, from conifers and pines which have been introduced into the country, is used extensively for building. In spite of their names some hardwoods, poplar for instance, can be quite soft and vice-versa.

Once a tree has been felled, as is commonly known, it is quite easy to establish the age by counting the number of growth rings, one for each year, which develop immediately under the bark. The outside few rings of sapwood are soft and immature while those further in form the more durable heartwood. A knot forms where branches leave the main trunk and is not usually a desirable feature.

The way a tree is cut up is of great importance. The easiest way is 'through and through' in which the wood is cut into planks of the required thickness. This method cannot be used when the grain produced by the annual growth rings does not run in the required direction. It will not do for making such items as cricket bats, barrel staves and furniture, or for carving wood, oak in particular, when it is desirable to show the medullary rays which grow outwards from the heart. Other methods can be used, quarter sawn for instance which is a more complex way of cutting and some wastage is accepted to obtain the required grain structure. The log is sawn into quarters which are then sawn individually so that the grain of the wood runs in the desired direction.

Seasoning

The best time to cut wood is in winter when the tree is dormant and therefore contains less sap. Newly cut wood contains a great deal of moisture which is normally allowed to dry out before use in a process called seasoning. Wood seasons at about an inch a year, and can be left 'in the round' or else cut up and seasoned 'in sticks' with small wooden blocks between each plank to allow air to circulate. Nowadays, most wood is seasoned in a kiln for about a fortnight to speed up the process and to dry the wood sufficiently to stand up to a centrally heated environment.

Not all trees are left to grow to maturity. Some are 'coppiced' or cut to the ground every eight years or so to encourage the growth of a number of shoots rather than one central trunk. This type of wood is needed for making hurdles, besom brooms and for chestnut fencing. 'Pollarding', where a tree is cut some eight feet from the ground, is not much practised today though it is sometimes seen in residential areas to prevent a tree becoming too large. In the past, trees were pollarded every twenty-five years or so to provide firewood and the replacement shoots grew out of reach of deer and rabbits which might feed on them. A good example of pollarded wood can be seen in the trees of Burnham Beeches.

Most wood is sawn, though the craftsman usually prefers to 'cleave' it by splitting it along the grain. This enables it to retain its maximum strength, and all thatching spars and hurdles have to be made of cleft wood.

Pit sawing. Tree trunks used to be sawn into planks over a pit. The pit was dug in the ground and the trunks rolled over it and held in place by metal dogs. The top sawyer stood on the trunk and the underdog worked in the pit. The cutting was done on the down stroke so the underdog got the sawdust all over him while the pleasanter job of the top sawyer was to pull the saw upwards and to keep the saw sharp. The job of pit sawing was uncomfortable and hard work but the system worked well and could be used today in underdeveloped countries lacking mechanical aids.

Tools

Tools are all-important to the rural craftsman who selects them with an expert eye and looks after them carefully. They are frequently made by a local blacksmith and so vary from individual to individual.

Bill hooks are used extensively for trimming, cutting and cleaving and must be of high-quality steel so that they can be kept razor-sharp. Varying shapes are favoured in different regions.

One of the woodland craftsman's most important devices is the brake, used to hold firm the material which is being worked. When the handle is pulled down, the 'jaws' of the brake are opened and the piece of wood to be worked is inserted. The lever is then released and the wood held in a vice-like grip due to the weight on the end of the chain. A draw knife can then be used to shape the stave, which is probably of wood cleft by a fromard.

Hedge-laying

Hedging and ditching were traditionally winter jobs when there was not much other work available. Hedges are laid by cutting away part of the stems of the hedgerow saplings and laying them over at an angle of 65° or more. The saplings grow in this position and the branches tiller out to form a thick network making the hedge stockproof.

To hold the hedge together and improve the appearance, some of the saplings can be left vertical and cut off at the top of the hedge. Alternatively, stakes can be driven in every two feet and a pair of hazel rods entwined between them along the top of the hedge. The craft of hedge-laying is now very much on the decline largely owing to labour costs. An electric fence is obviously just as effective and easier to erect although not perhaps so attractive.

Besom brooms

These are made by a broom squire from the twigs of birch that has been coppiced and cut every four or five years then left to season for a few months. The birch is made up into small bundles about thirty inches in length and bound together in the past with cane or bramble but now invariably with wire. The broom squire sits at his horse, applies the brake with his foot which clamps the wire enabling it to be firmly bound round the twigs. Any available wood is used for the handle, one end of which is pointed, forced into the broom head and secured in place with a nail or wooden peg. More rarely the brooms are made of ling heather but this is not readily obtainable in long enough lengths.

Hay rakes

The countryman always uses a wooden rake for gathering hay or leaves. These can be made in a variety of ways, but preferably of cleft wood and willow is admirable for the purpose. A piece about two inches square and thirty inches long is cleft for the head and shaped with a draw knife. Between eleven and fifteen holes are made in the head to take the teeth which are some four inches in length and are driven in to make a tight fit before the ends are sharpened and trimmed.

The handle can be of any suitable wood – the end is cut in a Y-shape and the two arms are driven into holes in the head at a slight angle so that head and teeth 'come home', inclining towards the user of the rake.

A similar type of rake can be made from sawn ash wood with handle and teeth of machine-made dowels. It is equally effective but lacks the character of the traditional cleft wood rake.

Hurdles

The wattle hurdle is made from coppiced hazel of about seven or eight years growth. A number of holes are bored into a slightly curved piece of wood called a 'mould' to take the nine or ten vertical posts or 'sails'. Hazel rods are cleft in half with a sharp bill hook, the only tool needed, woven in and out of the sails, then finished off with a careful twist and turn so that the wood fibres are not broken. The work is kept tight by pressing down with the foot, then the knee and finally by using the back of the bill hook. This type of hurdle was originally used for penning sheep but the majority are now used as garden screens.

The other type of hurdle is the bar or gate type, normally made of cleft ash trimmed to shape with a draw knife. A special tool is used for making the mortice holes and the gate is stiffened by nailing on cross bars. When used for penning sheep, this type of hurdle usually has six bars, those at the bottom spaced slightly closer together to prevent lambs getting through. They can also be used for horse jumps, often interlaced with birch, or as temporary gates and in these cases usually have four bars.

Oak gates

Farm gates vary in design from region to region but the standard shape consists of five bars, the lower ones closer together to prevent lambs squeezing through. At one time they were made of cleft wood but it is now invariably sawn. The gates are held together by wooden pins, coach bolts or nails clenched by turning over the end. These traditional oak gates, left unpainted, are sound for forty years and more. Their normal width of about ten feet is insufficient for a modern combine harvester, so they are gradually being replaced by wider metal gates.

Walking sticks

The making of walking sticks and shepherds' crooks is mostly a hobby or pastime practised over a wide area and taken very seriously in the Border country. Holly, hazel and ash is the wood most frequently used. In order to get the walking stick shape, saplings can be grown straight until they reach the right thickness and then the crook steamed to shape. Alternatively, a two or three year old ash sapling can be cut down to about six inches and replanted in a horizontal position. A branch will then grow vertically out of this so that after a few years the walking stick, complete with right-angle handle, can be cut and finished.

More simply, suitable sticks can be found in the hedgerows and the crook either tooled to shape or fitted with a handle. The ramshorn for the handle is boiled until pliable and then bent as required, smoothed and given a suitable finish, the wood frequently being coated with polyurethane to preserve it and accentuate the grain.

20

Charcoal burning

Charcoal is used today to make a surprisingly wide assortment of products ranging from gunpowder and fuel to dog biscuits and drawing material.

The traditional method of production was to make a 'clamp' which was laboriously built up around a central funnel with small sticks inside and larger ones outside. When complete, it was covered with a few inches of earth, ignited down the central funnel, sealed off and left to smoulder for several days during which time it had to be tended continuously. Should a gap occur which was not plugged immediately, the clamp would flare up and quickly be reduced to ashes. The old clamp has recently been superseded by the modern metal kiln. Two or three metal cylinders about eight feet in diameter are transported to the wood supply and filled with logs. The lid is put on firmly, the joints are sealed with clay and the wood is ignited and left for two or three days to be turned into charcoal.

Wood turning

At one time hand-turned bowls and wooden spoons were used for eating on an extensive scale and they are still produced in reduced quantities. Nowadays the lathes are power operated, the piece of wood to be turned being screwed to the revolving chuck and the bowl fashioned to shape with a chisel which is usually held against a support to keep it steady.

Most hardwoods are suitable for turning with a preference for walnut, elm, sycamore and beech. The wood is sandpapered to a fine finish and a treatment with oil or polish accentuates the grain structure.

Clothes pegs

Often made by gypsies, clothes pegs are cut from
pieces of wood such as willow or poplar and fixed with
a metal band round the middle. The 'mouth' is cut
afterwards with a knife.

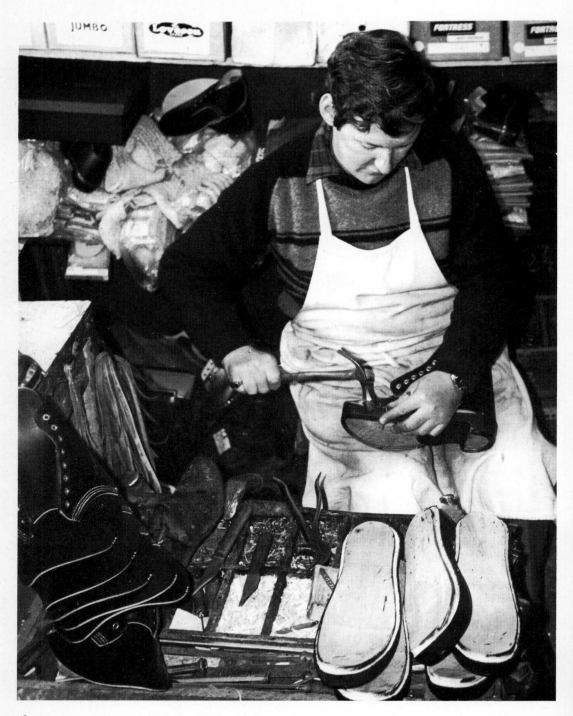

Clogs

Clogs are boots or shoes with leather uppers and wooden soles fitted with iron plates or rubbers. Watertight and warm, they are ideal for wearing on farms and factory floors. The soles used to be cut from the wood of the alder tree but these days beech is more common. The edge is cut away enabling a welt to be fitted flush with the sole and the leather uppers are then nailed to the soles. Clogs are comfortable as well as cheaper than leather-soled footwear and better for the feet than rubber and synthetic soles. They are still worn in Lancashire and Yorkshire and for clog dancing in which there appears to be renewed interest.

Baskets

Willow baskets. The majority of baskets are made out of willow rods or osiers. Suitable willows are grown over a wide area though most of the production comes from Somerset. The willows are cut to the ground every winter and send up shoots which grow 3–5ft during the summer. The bark is stripped from all but the roughest willow rods. These 'buff' rods are boiled and chemicals in the bark dye them their characteristic brown colour. Finally, the bark is stripped by machine and they are spread out to dry in the open air. Best for basket making are 'white' rods, the bark of which is stripped in spring when the sap starts to rise and it can be easily removed.

To make a basket, the rods have to be dampened to render them pliable. The basket-maker sits on the floor, in front of his sloping board with tools and his supply of rods beside him. The rods are intertwined in a variety of ways that keep the work strong, rigid and even. The basket makers are paid by the amount they produce and work with great speed and sureness of touch. Such baskets are used to transport animals or for picnics, and fishermen use the quarter cran baskets for measuring herring catches. The industry is declining, making such items as laundry baskets hard to obtain.

Sussex trugs. Made in the Hurstmonceux area of Sussex, these trugs are composed of strips of willow about an eighth of an inch thick and varying in length and width according to the size of the basket required. The frame or rim is shaped in chestnut made pliable by steaming, and tacked at right angles to this is another piece of chestnut for the handle which passes right underneath the body of the basket. The willow laths are shaped at the ends, steamed until pliable and held in position on the frame and handle with copper tacks. This type of basket is unsurpassed for garden work and lasts indefinitely if kept dry, but with few craftsmen left producing them, the supply is limited.

Spale baskets. These general-purpose baskets are made of interwoven laths of cleft oak. Though they used to be comparatively common and cheap, less than a dozen men are now making them in the Furness area of North Lancashire. Lengths of oak, approximately four feet long are boiled for several hours and cleft into laths about three inches wide and a sixteenth of an inch thick. The cleaving is started by using a tool called a froe and the lath is then peeled off by hand and trimmed by a plane or draw knife. A hazel rod about an inch in diameter is steamed and bent to form the rim and the laths are then fixed on, first those going across then the lengthwise ones which are woven in and out. The final ones are trimmed so the fit is exact and the basket can hold grain without it escaping. The laths are worked wet and are quite pliable and after completion the basket must be dried quickly to prevent mildew.

Rush baskets. These are of small and simple construction. Rushes are gathered in mid-summer and carefully dried. Using a block of wood of the required shape the rushes are woven in and out round it to form the basket. Alternatively, the rushes can be plaited into long lengths and the plaits wound round any object that will give the right shape. They are held in position by twine threaded through them with a long needle. The finished article is very suitable for use as a log basket.

Pole lathes

Cheap and effective, these implements were used by chair-bodgers for turning legs until around twenty years ago. The main function of the pole, frequently a rough bough, was to keep turning the piece of wood as it was being shaped by the turner's chisel. Held in position by wooden uprights, the horizontal pole was connected to the lathe by a strong line which continued down to the foot-operated treadle. When the treadle was depressed, the string wound down turning the wood towards the operator, and when pressure was not applied the bough, bent over like a fishing rod, sprang back into position, turning the wood the other way.

33

Coopering

Although metal barrels are now most commonly used there is still a demand for wooden barrels for beer and maturing whisky. English barrels are made exclusively of oak but, curiously, English oak is not popular. The most sought after variety, from Memel in the Baltic, is no longer available so that oak has to be imported from other countries. A cask is made of a number of staves bound together with iron hoops. There are circular heads at either end and holes for the bung and tap. Each stave has to be carefully tapered and hollowed with the sides cut with an angle to form the circle of the barrel. Every task is done by eye using tools peculiar to the trade. For 'trussing up', the staves are either steamed or heated over a fire to make them pliable so they can be forced into shape when the hoops are put on. A special plane called a croze is used to cut out a groove inside the top and bottom of the barrel to take the heads. The heads are made watertight by packing with rushes but the staves are fashioned with such precision that no packing is necessary between them. A well maintained barrel can last for some fifty years.

Coracles

Coracle design has remained unchanged for thousands of years. A framework of thin willow laths is interwoven to form the body with a gunwhale of woven willow rods. The cover, traditionally skin and pitch, is nowadays made of calico with a coat of pitch mixed with linseed oil giving a finish like shiny rubber.

The coracles still working operate in pairs, usually at night, on two Welsh rivers using a small net between them for catching salmon and sea trout. They drift down-river, blunt end first, kept apart by their single oar, and the net is quickly hauled in by the master boat as soon as a fish strikes it. Having drifted a mile or so down river the fishermen pull into the river bank, sling the coracle, weighing about thirty pounds, over their shoulders, and carry it up-stream to repeat the operation. Licences to fish using a coracle are not being renewed so the boats will soon become museum pieces.

Cricket bats and balls

Originally, cricket bats were made by a local woodworker and although the craft has become progressively more sophisticated, it is surprising how little the materials have changed over the years. The blade is made from a special species of willow which is now scientifically grown like a farm crop, and the handle, which is spliced into the blade, is of imported Sarawak cane with rubber springs, bound with thread and fitted with a rubber handle.

The willows are large enough to cut when they are about sixteen years old so demand has to be anticipated well ahead. Blades throughout the cricketing world are made of English willow (apart from a few from Pakistan) and nothing has yet proved superior. The weight of the bat can vary although width and height must conform to certain regulations, but even with the initial shaping of the blade and splice done using a jig, the finishing is by hand so no two bats are exactly alike.

For important matches, only cricket balls produced by hand in the old way are considered good enough. The core is a small cube of cork around which is wound wet worsted thread, thin pieces of cork being added as the winding continues. With the help of hammering in a brass cup the core is made perfectly round and as the worsted dries it binds tighter to give the ball its bounce. When the core has reached the required diameter the leather cover is put on. The best covers are made of four segments with pairs of segments sewn together on the inside so the stitches do not show, and the two halves stitched together with three rows of (exposed) stitching to form the seam. Finished balls must be exact in weight and size and over 100,000 are turned out annually in the Tonbridge area of Kent.

Golf clubs

Since the coming of the steel shaft in the 1930s, few individuals are left hand-making golf clubs. An exception is Tom Auchterlonie's workshop in St Andrews where wooden-headed clubs are still made. The steel shafts and roughly shaped heads are machine made separately. The heads used to be made out of persimmon wood which is now difficult to obtain and a substitute has been found in laminated maple. The roughly shaped heads have to be trimmed using a file and rasps and the weight is adjusted by letting in pieces of lead under the brass sole plate. The player can buy a club tailor-made for weight and balance, giving him the absolute confidence in his equipment so necessary for success.

Fishing rods

Over the years, fishing rods have become very sophisticated pieces of equipment, meticulously made and perfectly balanced. The majority are now made of glass fibre or carbon fibre though many connoisseurs prefer to stick to the traditional split cane type.

The cane comes from the East Indies in lengths about twelve feet long and two inches in diameter. This is cut in three to five feet lengths, the pieces are split, the joint marks sandpapered smooth and finally baked to dry them out and temper the wood. Split cane rods are made in a hexagonal shape out of six pieces of cane which are triangular in cross section, tapering towards the end and glued together. A rod is made in two or three sections, the final one tapering to less than an eighth of an inch in diameter so they have to be made with great precision and care. The cane is either cut on a machine or shaped on a jig. The sections are joined together with waterproof glue, bound and hung up to dry. Finally, the rod can be varnished and the fittings added. The finished rods are extremely light, but have an extraordinary degree of flexibility, balance and strength.

Longbows

The longbow was developed by the English and brought them victory in the battles of the Agincourt campaign due to the fact that it could be shot at five times the speed of the French crossbow. Imported Spanish yew was used as English yew is too full of knots and blemishes. Bows were usually a little over six feet in length and skilfully cut from the main trunk of a tree so that the outside or back was of sapwood and the belly, the side towards the archer, was of heartwood. When the bow was drawn the sapwood was extended and the heartwood compressed to give the most effective performance.

The steel tipped arrow was made of scots pine or ash and could be fired over 200 yards. The ends of the bow had horn nocks with grooves to take the string, usually made of hemp or flax. The longbow, was superseded when firearms became effective but archery has remained a sport of kings and to this day the royal bodyguard in Scotland is the Royal Society of Archers. Today, bows are still being made for the increasing number of people who are taking up archery for recreation.

Iron

The blacksmith and wheelwright used to play a vital part in village life. Between them, they supplied the agricultural, household and craft equipment essential for everyone else in the community to carry out their work. The traditional blacksmith was responsible for repairing farm machinery, forging metal for wagons and gates, and for shoeing horses. His workshop and forge would have been situated near the heart of every village. In recent years, the craft has split so that the wrought iron worker or smith and the farrier each have their own specialised area of work.

The smith

Making new work or repairing old, the smith handles gates, fire dogs, weather vanes and other articles, both decorative and functional, and invariably makes his own tools for the job as well. Curiously enough, he works in mild steel which is wrought to shape by heating in the forge and hammering on the anvil. Wrought iron as such is not produced by modern smelting methods.

The farrier

The method of work of a farrier has undergone a complete change over the last few years. He still has a forge to make and shape the horseshoes but as a general rule few horses come to the smithy. Instead he sets off at dawn in his car loaded up with shoes and goes himself to horses on farms and in stables scattered around the countryside. A horse's hoof has an outside covering of horn that grows at the rate of about half an inch a

month. When grazing in a field shoes are unnecessary but for hunting or roadwork they must be fitted to prevent the hoof being worn down and causing lameness. A set of new shoes has to be fitted at monthly intervals firstly because the metal wears away but also because the hoof grows outwards and the old shoe begins to pinch. To work on a shoe, the farrier stands with his back to the horse with its hoof held between the knees. The horn is trimmed with a knife and the shoe tested

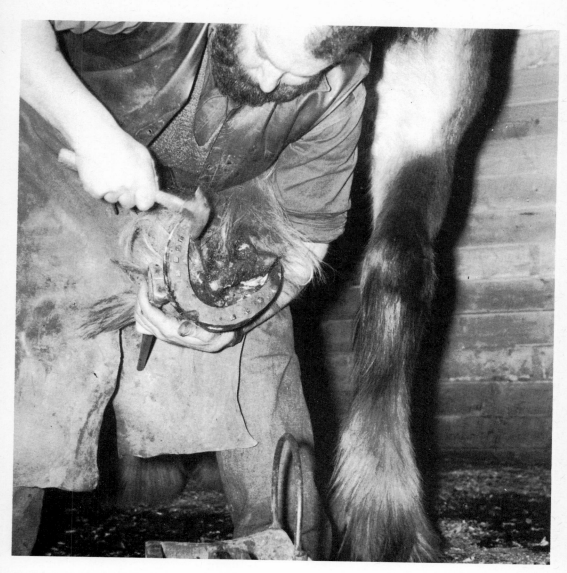

for shape and size. Nowadays, about eighty per cent of shoeing is done cold though a better fit is obtained with a hot shoe, cooled slightly in water before being nailed on. When placed on the hoof it sizzles and gives off clouds of smoke but causes no pain or distress to the normal horse though it can be frightening for a highly strung race horse. Specially shaped nails are driven through holes in the shoe and emerge on the outside of the hoof where the end is snipped off and the nail clenched over and filed smooth. Normally, four nails are used on the outside of the shoe and three on the inside. They have to be positioned with extreme care to avoid damaging the hoof or penetrating the tender part. With a large increase in the horse population over the last twenty years farriers are much in demand and are doing very well.

The wheelwright

Due to the high cost of petrol, people are once again beginning to get out their pony traps and to use horses for work as well as pleasure. Though there are only a handful of wheelwrights left they are again in demand for the making and repairing of wheels. The metal 'bond' or tyre for instance may wear out or come off if the wood of the wheel is allowed to dry and shrink. If a spoke breaks, the rim of the wheel has to be entirely dismantled – it is interlocked with 'dowels' and pieces cannot be removed individually. The tyres of most wheels are continuous rings although rarely they consist of 'strakes' or metal segments nailed round the rim. To be fitted, these tyres are heated up till nearly red hot to expand the metal, then placed over the rim of the wheel, hammered on and quickly doused with water to prevent setting fire to the woodwork. On shrinking, the tyre binds the wheel tight and helps to give it an additional 'dish'. The spokes are angled outwards slightly to increase the dish, so enabling the wheel to stand up to the inevitable sideways battering it will receive as the wagon is drawn along.

Earth and Stone

Stone

Stone is still a prestige material in the construction industry. In the past, buildings were of solid stone but nowadays the stone is probably only a few inches thick and is used to face the steel and concrete framework. It is also needed for repair work, particularly of churches. Stone varies enormously in hardness and colour from limestone to sandstone and granite. It may be quarried by digging from the surface, or mined as,

for instance, in the case of Bath stone. The cutting has often been mechanised with pneumatic drilling though there are many craftsmen still working away on a smaller scale. They split the stone by driving in wedges and using a long lever to snap off pieces. These are trimmed to shape and cut with a carborundum wheel or diamond saw. Freestone so called because it can be cut in any direction, can be cut by hand using a large saw called a frigbob.

Stone roofs

Some stone splits into thin layers and can be used for roofing tiles, found particularly in the Cotswolds. Suitable stone is quarried in the winter and left out to be frosted. The stone contains around ten per cent of 'quarry sap' which causes the tile to flake when it freezes. After drying out, the stone hardens, developing an outer crust that does not readily re-absorb moisture, so can be used for roofing without fear of further splitting. The stones vary greatly in size, the larger ones are used lower on the roof and sizes diminish towards the ridge. A hole is made in the top of each stone, usually with a power drill, the nail pushed through and the stone is hung from the batten, not actually fixed to it. The nails are covered as work proceeds and the tiles are kept in place by the weight of each succeeding course. The old method was to use oak pegs but nowadays copper is used with aluminium or galvanised nails as a cheaper substitute.

Slates

Requiring a great deal of labour to produce, slates have become increasingly expensive so that they can no longer compete with modern tiles. Slate is found mainly in Cornwall, Wales and the Lake District with regional variation in colour. The slate is quarried in large slabs, usually obtained by blasting, and cut mechanically into a size which is easier to handle. The craftsman who makes the roofing slates sits in a low chair with the slate block propped against his knee. The chisel is positioned carefully and struck with a 'beetle'. A few blows cause the slate to flake off, later to be trimmed, squared and eventually nailed to the roof battens. Each successive course of slates covers the nails below and the thin, level-surfaced slates make an excellent roof, waterproof and light.

Dry-stone walls

These walls are made without the use of any mortar and maintain their shape due to the skilful way they are built. They are seen at their best in limestone areas, in particular in the Cotswold country where the stone splits into shapes that are comparatively easy to work with a little help from a trimming hammer. A ditch about a foot deep has first to be dug, in which are placed the largest stones. The wall slopes in towards the top and is finished with vertical capping stones. The dimensions of the wall are gauged by a wooden template giving the cross section of the wall, and strings run along to give the builders guide lines. The outside faces of the wall are flush but the inside is filled with rubble stone, and 'through' stones, placed at regular intervals, key the wall together.

In Scotland these walls are called dykes and are sometimes made from the stones removed from the fields. In general they are more rounded in shape and pose greater problems to the wall-maker. Stone walls are normally about five feet in height and last indefinitely if properly maintained.

The centres of the Devon type of wall are normally filled with earth and they are then finished with a covering of turf along the top.

Hand made bricks

Brickworks have to be situated in the immediate vicinity of a good supply of the raw material, clay. The process of mechanisation producing cheaper bricks has caused the closure of nearly all the small firms making them by hand, and sadly so since the mass-produced article lacks the variation in texture and colour of handmade bricks. The clay for the bricks is 'pugged', a process in which it is mixed to an even consistency and stones and lumps are removed. The brickmaker then throws a lump of clay into a mould made of wood or metal and cuts off the surplus from the top with a wire stretched across a bow. The brick is then left to dry when it can lose about an eighth of its size in shrinkage. Finally, the bricks are stacked in a kiln and 'fired' by subjecting them to increasing degrees of heat.

The colour of the brick varies with the chemical constituents of the clay and the heat of the kiln. The greater the firing heat, the darker will be the colour, ranging from white through red and purple to black, and the clay can even be melted if the heat is too fierce. Variations in colour can also be achieved by adding chemicals during the pugging stage and by rolling the clay in sand before it is moulded.

Pottery

True country potters who were apprenticed to the trade are now very rare although there are plenty of artist-trained potters. The former make such items as bread crocks, chimney and garden pots and use about a ton of clay every day. Their potteries are situated near suitable clay which they dig and pug themselves. A piece of clay is then weighed out and 'wedged' by slamming it down on the bench to eliminate air bubbles. It can then be 'thrown' on a wheel which may be driven by foot though more often mechanically turned. The hands are kept wet and the clay is first centred on the wheel. The sides are raised and the pot begins to take shape. A big pot can contain twenty-five pounds of clay but as the potter has to reach down inside to raise the sides, the pot can only be as tall as the length of the arm.

After the preparatory work, a large pot can be made in about five minutes on the wheel. Afterwards it has to be dried slowly and later fired in a kiln, the grey clay emerging a red colour and slightly porous. Articles which are to be glazed have to be fired a second time.

Reed and Straw

Thatch

Thatching is historically one of the earliest types of roofing and is still used extensively today, especially with the current boom in converting old cottages. The good insulating qualities of thatch make it warm in winter and cool in summer. The disadvantages are its limited life, overhanging eaves which reduce the light in upper rooms, the increased insurance premiums on both the house structure and contents and the ever-increasing cost of renewal. There are thought to be around 50,000 thatched dwellings in Britain and 600 thatchers. Good thatchers are usually booked up for a year or more in advance.

The materials most commonly used are reed, most of which comes from Norfolk, and wheat straw. Rye straw is also excellent but is rarely grown nowadays. Ling heather is another long-lasting thatching material cut in lengths of about a yard. It was used extensively in the Western Isles of Scotland for roofing the crofters' cottages.

Norfolk reed. The most expensive and probably the best thatch is Norfolk reed, so called because most of it comes from that area although the reed grows in any brackish water. It grows 5–7ft tall and the supply does not meet the demand. The roof for thatching is stripped bare and layers of reed are laid on the battens. A rod of hazel about six feet in length is laid across the layer and held down by metal barge hooks hammered into the rafters. Each succeeding layer covers the rods, the reeds being beaten up tight by a tool called a leggett as work progresses.

As reed will not bend, moistened rush has to be used for the roof ridge, bent over the top and pegged down with hazel spars. Gutters are never fitted so thatch roofs have a considerable overhang to throw the rainwater clear of the walls. A Norfolk reed roof can last up to seventy-five years if it is maintained in good order and the ridge renewed periodically.

Wheat reed and long straw. Straw for thatching has to be specially harvested with a reaper and binder and then threshed using a reeding attachment. For preference a type of wheat with a long and hollow stem is used and artificial fertiliser kept to a minimum as it can cause brittleness.

There are two main ways of thatching using wheat straw and individual thatchers also vary these methods slightly in detail. Wheat reed or Devon reed looks the best and lasts longest, up to twenty-five years. For a newly thatched roof a layer of straw some nine inches in thickness is tied with tarred hemp rope to the roof battens. A top layer of another nine inches of straw is placed over this and beaten up tight with a leggett. The thatch is held down by hazel spars two and a half feet in length. When recovering an existing thatched roof, a certain amount of the old straw is pulled out and one nine inch layer of new straw is added on top.

Using the method called long straw thatching, the top layer of thatch is put on more loosely and is not beaten up tight. A pattern of spars, usually criss-cross, always shows along the lower edge of the roof. This type of thatch has to be wired over to prevent damage from birds. Some thatchers finish off with straw ornaments for decoration.

In the case of Norfolk reed and wheat reed the brunt of the elements is taken on the end of the stems and an inch or two above but with long straw, rain impinges more on the stem. As a result, this type of thatch only lasts for about fifteen years and is gradually falling out of favour.

Thatching spars. Hundreds of thatching spars are
needed for every roof thatched with wheat straw.
They are made of hazel, preferably pieces one inch in
diameter which are split into about four lengths of two
and a half feet. Spars are laid across the straw and held
down by other spars, twisted into a hairpin shape,
which are pushed down tight into the thatch. Spars can
be made by the thatcher on a wet or windy day when
thatching is not possible, or they can be bought from a
private supplier.

Rush and cane chair seats

Rush for seats comes from the common bulrush which has to be cut and dried in midsummer while it is still pliable. After dampening, three rushes are twisted into a rope and woven around the chair seat starting at the edges and working inwards. New rushes are twisted in to make the work continuous and finished off on the underside of the seat.

Cane comes from the East Indies and is cut in lengths several yards long. It is supplied in a variety of widths and is interwoven in a number of patterns which give an attractive criss-cross finish with the shiny part of the cane showing uppermost. The work is slow and meticulous and was much favoured for the seat and backs of furniture from the eighteenth century onwards.

Corn dollies

This ancient craft spread from the Mediterranean. Of pagan origins, corn dollies were believed to keep the spirit of the corn alive till the next sowing season. Usually they are made of wheat straw of good length and colour with a hollow stem, and in a number of traditional patterns. The straw must not be squashed flat and is often worked evenly round a core of straw or wire. Dollies can last for years and are now made chiefly for decorative purposes.

Wool and Fibre

Wool

A sheep's coat grows through the summer and in the natural state is shed in a moult every year. The wool of domestic sheep is sheared off in spring or early summer. The quality and length varies with the breed and condition of the sheep, and the part of the body from which it comes. Originally, all shearing was by hand and on occasions is still done this way for trimming the coat for show, or where only one or two animals need shearing.

A good shearer removes the wool in one piece or 'fleece' which is folded and sent to the Wool Marketing Board to be graded by a 'stapler' who then sells to individual weavers.

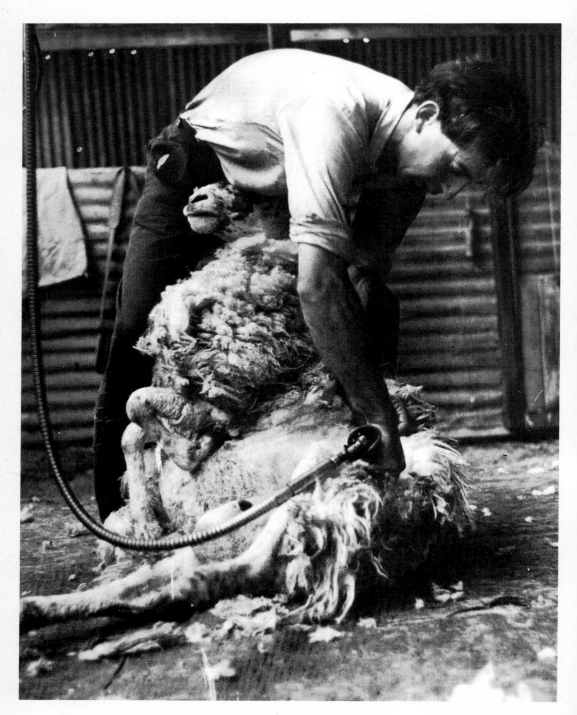

Spinning. The wool is first 'carded' by rubbing between a pair of hand cards. These consist of a large number of wire combs which get the hairs running in the same direction and remove some of the dirt, a process that is usually carried out before washing while the wool still has its natural grease. Little if any wool is hand-spun for a living as the job is slow and laborious. However people are taking up spinning as a hobby on an increasing scale and spinning wheels are still being made commercially in significant numbers.

The simplest way of spinning is to use a distaff and a spindle on which to wind the thread. The wool is fed by the left hand and spun on to the spindle with the right. The spinning wheel is operated with the foot which causes a large grooved wheel to revolve. From this, two strands of string drive the flier mechanism and the bobbin which both run on the same spindle. Wool is fed into a hole in the end of the spindle and emerges out of the side, passes round one of the hooks, called hecks, on the flier and is then wound to the bobbin. Each time the fast moving flier revolves, one twist is put in the yarn. Wool must be fed in very carefully to obtain yarn of an even thickness. Yarns are woven together to form the required ply either for knitting or weaving.

Hand-weaving. The warp yarn, which runs lengthwise, is wound onto the warping drum. The other ends are passed through the heddles and attached to the cloth drum. The heddles, of which there may be several, are like combs which can be raised by a foot pedal forming a gap or shed through which a shuttle can be pushed with one crosswise thread or weft. A beater is pulled towards the operator after every weft thread to keep the pattern tight. By alternating the lifting of the heddles and the colour of the weft threads a variety of different patterns can be woven. Hand-weaving is gaining popularity as a vocation although the number of craftsmen hand-weaving for a living is limited to a very few still left in the outlying islands of Scotland.

Quilts

Quilts for beds have been used for several hundred years both for their decorative effect and their warmth. Normally, there is a patterned front and a plain back with a layer of wool or cotton in between to give warmth. Traditionally, the work was carried out at home, hand-stitched right through from front to back, and the top occasionally built up in a patchwork design. A single cover might take a fortnight's work.

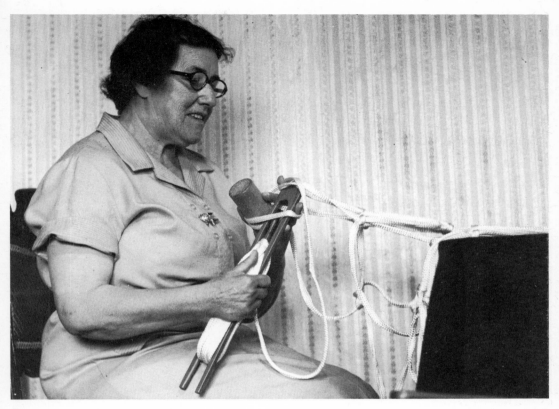

Lace

Hand lace making has always been a cottage industry to provide a little extra money in the home. Progress is slow and the returns small so that those still left lace making usually do it as a hobby. The craft was brought to this country by refugees from France and the Netherlands and patterns vary under their differing influences. In pillow lace making a pattern of cardboard or parchment pricked with numerous holes is fixed to a pillow stuffed with hay or flock. Linen thread, or possibly cotton or silk, is wound round bobbins which are worked in pairs, one by each hand. The threads are crossed over one another to form the pattern with new bobbins brought into play to increase the complexity. The number of bobbins varies from a few dozen to over a hundred for a large and complicated pattern. Special brass pins are placed in the pattern to keep the stitches in place, which are moved from back to front as the pattern progresses towards the lace-maker, while the pillow is revolved to keep the work central.

Net

Net was usually made of hemp, tarred when used for fishing, though other materials such as cotton were also used. These days the majority of nets are machine-made of artificial fibre. Machines still cannot cope with nets not of standard dimensions or those in which the mesh gets progressively smaller – the pockets of billiard tables for example – and around Bridport, Dorset, the work is sent out to the villages where women have done this type of work for generations.

Apart from the netting material, all that is needed is a mesh stick cylinder to gauge the size of the mesh and a 'needle' specially fashioned out of hardwood – the example in the photograph is apple wood. The knots are made by a few deft movements of the needle and, paid by results, the operators work swiftly and surely. Deep-sea fisherman who overhaul their nets on arrival in harbour use the same kind of netting needles.

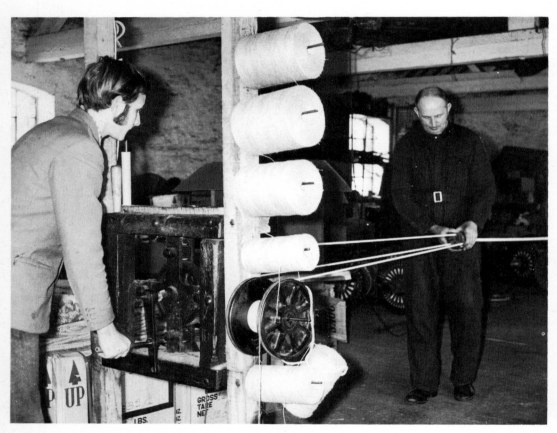

Rope

At one time there were numerous 'rope walks' dotted about the country producing hand-made hemp rope. The small quantity of rope still produced by hand is usually for specialised purposes, for show animals and church bells for instance.

A 'jack' fitted with a handle rotates three hooks spread in an eight inch triangle. Long strands of the rope fibre are fixed to the hooks and secured at the other end to a weighted platform or 'traveller'. The rope-maker inserts a piece of wood or 'top' to separate the strands at the traveller end and his assistant turns the jack wheel. Then the strands begin to twist and the rope-maker walks toward the jack holding the top as the rope forms behind him. The rope shortens as the strands twist so that the traveller is towed along behind. When the top arrives at the jack, the strands are unhooked, secured and the rope 'back twisted' to prevent kinking.

Hide and Horn

Tanning

This is the process which turns animal hide into leather after the hair is removed, preserving the skin and keeping it supple. No successful substitute has yet been found for leather and enormous quantities are required for footwear and a variety of other purposes. Hide is tanned by soaking it in a solution of a chemical found in the bark of trees. Traditionally, it was the tannin of oak bark, but nowadays chrome is used as it works more rapidly. The tanning process takes several months, during which time the complete hides are moved to progressively stronger solutions of the chemical.

Saddlery

The explosion in the horse population since the war has kept saddlers busier than they have ever been. Every horse should have its own specially made saddle which will last well over fifty years (if well maintained). Nearly all the work must be done by hand including the cutting out and all the important stitching. The breakage of a vital piece such as a stirrup leather could cause a bad accident so only the best leather is good enough. Saddles are built on beechwood saddle trees and vary from the substantially made hunting pattern to the (small, light) racing saddle. The horse population may well have reached its peak, partly due to the high cost of feeding stuff but there is little need for saddlers to worry about a work shortage. This country still produces saddles at a lower price than abroad so there is plenty of scope for export.

Harness

Saddlers tended to specialise in making saddles or harness, the latter frequently referred to as black saddlery. Sometimes a further distinction was made between making horse collars and other types of harness. There has been an upsurge in demand for horses to work on the land, to pull brewers drays, for coaching contests and for private enjoyment which has led to a new demand for harness just as the industry was almost dead.

The details of harness vary considerably depending on the number of horses pulling a vehicle and the arrangements of the wooden shafts. The collar is the most intricate piece of equipment stuffed with rye straw if available, and wool flock. Each horse should have its own and preferably a second one and it is essential that it should fit correctly or it will cause soreness and obstruct the windpipe. The straps of harness are nearly all adjustable with buckles and are frequently decorated with brasses.

Gloves

Glovemaking is largely a cottage industry with firms employing outworkers who make the gloves at home. Gloves can be machine-stitched, but the best work is done by hand using high-quality materials. The cottage worker is provided with ready cut tranks forming the main part of the glove which have been cut by the webber. Pieces are supplied to form the thumb and sides of the fingers and these are sewn together, frequently with a lining. Although most gloves are made of leather, silk, cotton and other materials can also be used. Much of the best skins come from sheep specially bred for glove leather, mostly from Abyssinia or the Cape.

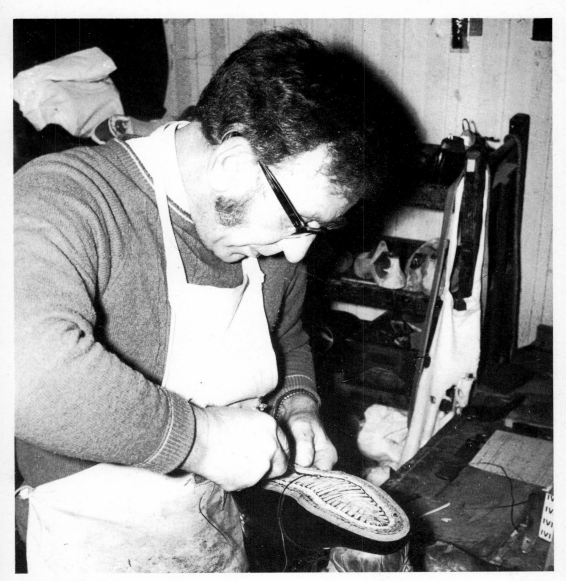

Boot making

The successful mechanisation of shoemaking is a comparatively recent development. Up till the beginning of the century, nearly all footwear was hand made and every village had its own shoemaker. Now very few individual bootmakers are left and they supply the luxury market and special surgical footwear. The shoes are worked on a wooden last, individually fashioned to the shape of the customer's foot. All the important pieces are hand-stitched with best hemp thread that the craftsman makes up himself to the required thickness. To make shoes by hand is considerably more expensive than mass production but the customer has the satisfaction of knowing that his shoes are of top quality leather and that the fit will be perfect.

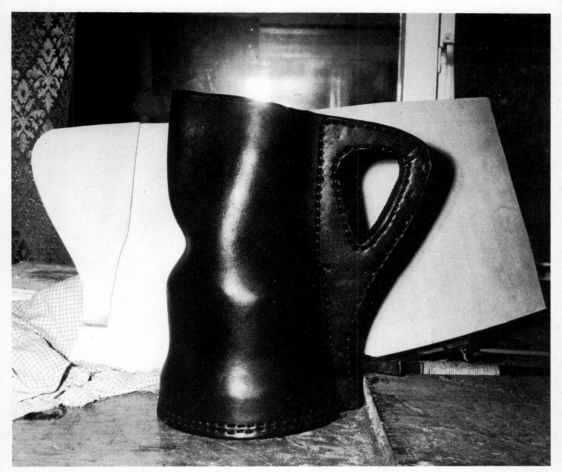

Parchment and vellum

Though the demand for vellum and parchment has declined it is still used for documents such as presentation addresses, for drums, or for book covers. Vellum is made from the skins of young calves and goats and is light brown in colour whereas parchment comes from the inside of split sheepskin and is pearly cream. The skins are soaked in a special chemical after which the hair is removed and they can be stretched tight on a wooden frame and left to dry. The skins are then shaved by a specially shaped knife to remove fat and unwanted surface impurities. Finally, they are rubbed down with pumice and treated with French chalk.

Bottels and jacks

Leather was most popular for drinking vessels in medieval times when it was plentiful and cheaper than the fragile and expensive glass or pewter which were also used. There are a few craftsmen left making these leather drinking vessels – bottels, which are barrel-shaped with a short neck in the centre to take the cork, and jacks which are large jugs. The outline is cut from a pattern out of one piece of leather, then soaked in water, formed round a wooden block and clamped in position. When the leather is dry, the blocks are removed and the pieces hand-stitched together. The leather is dyed black and finally hot pitch is poured in and out leaving a thin layer around the inside of the vessel to make it watertight and easy to wash out.

Bagpipes

The bag of this instrument is made of hand-stitched sheepskin leather so remains slightly porous. The musical parts – three drones, inflating tube and chanter – are let into the bag which is then covered in tartan. Northumbrian pipes differ slightly since the bag is blown up by bellows worked by the arm. The only mechanised part of the process is the turning of the wood for the fittings which are made of African hardwood while supplies are obtainable, and of silver, ivory or plastic.

Taxidermy

Shooting enthusiasts, fishermen and ornithologists have always wanted to preserve examples of their trophies – the taxidermy firm of Rowland Ward has been established for over 150 years. Specialised skills are called for; the legs of birds and animals are reinforced with wire or steel rods bent to shape and the body is made on a wooden frame with stuffing and plaster. As a rule, the only part of the animal's skeleton to be retained is the skull complete with teeth, and in the case of a bird, the skull and beak. The skin is preserved, cured, and then soaked in water and formed around the framework that has already been made. It is then invisibly stitched up and the glass eyes, painted on the inside, are inserted. In the case of fish, plaster casts are made and painted to form an exact and faithful copy of the original.

Falconry furniture

Practised for thousands of years, this sport is once again on the increase. A variety of hand-made equipment is needed including a glove to protect the arm from the falcon's sharp talons and a hood to shield the highly strung bird from the goings on around it. When the bird is required to fly, the hood is slipped off.

The hoods are cut from leather in a variety of shapes and colours and hand-stitched. Decorative as well as functional, the hoods were traditionally topped with heron's feathers but any attractively coloured feathers are satisfactory. Working hoods are less ornate and are usually fitted with thongs instead of feathers.

Horn

Horns from cows are hollow until approaching the tip and this has made them very useful as drinking cups and powder horns. Cut and then heated, horn can be flattened and used in windows as a substitute for glass or cut out as combs. Pressed under heat, it can be used to make spoons and shoe horns. A steady amount of hornwork is still made though most of the material has to be imported as the majority of cattle in Britain are now de-horned.

Deer horn and ramshorn are nearly solid and these have their uses as knife handles and walking stick handles. If necessary they can be made pliable by boiling then clamped in a vice to retain the required shape.

Wind and Water

Water-mills

Water has long been used for providing power; and over 5,500 water mills are recorded in the Domesday Book. The majority were for grinding corn, the rest for pumping water or driving machinery. It goes without saying that they needed to be close to a water supply that did not fail and the type of wheel fitted depended on the height of the head of water. If necessary, this could be adjusted artificially by building a weir across a river or by regulating the water in a mill pond by sluices. A few coastal mills were worked by the tide. The least efficient and most primitive type of wheel is the undershot wheel fitted with paddles and driven round as the water passes by. This is only adequate if there is a plentiful water supply. More efficient is the breast shot wheel in which the water is fed in half way up the wheel. The water pours into troughs, the wheel revolves and the water is emptied out at the bottom. Better still is the overshot type in which the water is fed in over the top of the wheel. Less water but a higher head are required. Most effective of all are water turbines which are usually fitted beneath the mill and cannot be seen.

Windmills

Windmills came into their own where there was an insufficient supply of water or where the flow was too sluggish to use a water wheel. The earliest type was the post mill in which the whole wooden structure could be moved round by a tail pole until the sails pointed towards the wind. The massive quarter bars that support the structure can frequently be seen under the main body though some mills have a round house encompassing the base at ground level to give storage and protect the beams. Post mills were superseded by smock mills in which only the cap and sails revolved. The main structure became larger and more solid and was of wood with boarding on the outside painted black or white to preserve it. The cap was turned into the wind by a clever piece of mechanism called the fantail which saved the miller a great deal of time by doing this job automatically.

The final development along the same lines as the smock mill was the tower mill. It was more solid than the others, built of brick, sometimes with five floors.

The sails. The first sails were of canvas stretched on a wooden framework that had a slight twist like an aeroplane propeller. The number of sails varied from two to eight. Later the sails were made of wooden louvres that could be operated by springs, and finally patent sails were used that could be adjusted with the sails still in motion.

The machinery. The internal machinery of water and wind mills was basically the same. The drive came through wheels fitted with wooden teeth. These were later replaced by cast iron though some cast-iron wheels had wooden teeth which could be replaced more simply in the event of an accident.

A water wheel drove a pit wheel which in turn revolved a wallower. The wallower was fixed in a horizontal position on the main drive shaft which was itself set vertically in the mill. In a windmill, the wallower was driven by a large brake wheel which had an adjustable band around its circumference acting as a brake. The main shaft had a drive for hoisting sacks to the top floor so the grain could be fed by gravity to the revolving stones that crushed the grain, and these stones were driven by geared wheels from the main shaft. Stones were always used in pairs, the bottom bed stone remaining stationary and the top or runner stone revolving. Stones were usually made of millstone grit from Derbyshire, or French burr. They had grooves cut in them and as the grain was fed into the centre of the upper stone it was sheared and ground by the furrows cut in the stones. These furrows had to be recut every few weeks to keep the edges sharp. The flour worked its way out to the edges of the stones which were surrounded by a wooden casing and dropped down through a pipe to be collected in a sack below having become quite warm in the grinding process.

Water divining

Though not strictly a craft, water divining deserves a place in this book as an art long and effectively practised by countrymen. Inevitably, it is largely superseded by mechanical methods of finding water, or made unnecessary by modern methods of irrigation. These methods may be more certain but the curious mystique of the water diviner will be missed.

The divining twig can be made of a variety of different materials. A piece of whalebone or two wires can be used although a Y-shaped twig of hazel or elm is more usual. The diviner holds the two arms of the Y and walks slowly until he passes over an underground supply of water, when the twig bends and possibly breaks and the diviner's whole body may shake. The ability to 'dowse' is a special gift and is often only discovered accidentally. Many water diviners even claim that they can estimate the depth and strength of the water-flow by the strength of the pull exerted by the twig and the area over which it operates.

Dew ponds

These were ponds to catch rainwater and were dotted over the downs where there were flocks of sheep in need of drinking water. They consisted of a saucer shaped depression with a layer of consolidated puddled or kneaded clay two or three inches thick which was watertight. Over this was a layer of lime to set the clay hard and prevent worms penetrating, then sometimes a layer of straw and finally a foot of earth to protect the clay. The ponds were filled by rainwater and the water lost through evaporation was made up by dew. The same principle of puddling clay was used to make the beds of canals watertight.

Water meadows

Flat fields near rivers were frequently turned into water meadows which were drowned in the winter months by allowing water to flow over them so that they kept warmer than the surrounding frozen ground. In spring the grass grew sooner giving an early bite for the sheep. The meadows are looked after by a drowner who has cutting tools to keep the water channels free of weeds. A few acres of water meadow still exist in Wiltshire.

Photographic details

All the photographs have been taken during the last seven or eight years with Mamiyaflex C.3 cameras taking 2¼in square pictures using Kodak Tri X film size 120 processed with Unitol developer. I have used the 2¼in size instead of 35mm as I prefer the bigger size for making exhibition prints of 20in × 16in. I work with two cameras, one loaded with Tri X film and the other with Kodak Ektachrome X for colour photographs to illustrate lectures. The lenses used are 65mm, 80mm and 110mm.

For inside work, natural light is used if possible and the camera is hand held down to 1/30 sec. Electronic flash has been used extensively for much of the work. Using two cameras, several lenses and a flashlight unit at the same time taking notes, changing films, etc, means that I try and avoid using a tripod though always have one at hand.

Very little posing has been done as it invariably shows in the photograph and leads to frozen looks. My aim is to take the craftsman in his natural surroundings concentrating on the work in hand.

Appendix   Where to see crafts and craftsmen

The information in this book is the result of a lifelong interest in crafts which, over the last ten years in particular, has led me to record in detail, both in words and pictures, as many of the old craftsmen as I can find in my travels throughout the British Isles. To those who have a similar interest, my main advice would be, not simply to visit the many collections of old tools and examples of crafts which are now assembled in various parts of the country – excellent and informative as the majority of these rural museums are – but to get out in to the country as much as possible and try to see the remaining craftsmen at work.

Throughout the text I have mentioned one or two craft companies still operating, such as Tom Auchterlonie's golf club workshop at St Andrews, Scotland, and the staff of such firms are usually quite happy to demonstrate their art to the genuinely interested. But I would not recommend just turning up on impulse and demanding an audience as it is always best to write and ask for an appointment. The Forestry Commission's Maywood Centre near Warwick has demonstrations of the old woodland crafts such as the pit sawing and turning described on pages 8 and 24 and, of course, the Council for Small Industries in Rural Areas (CoSIRA) at 35 Camp Road, Wimbledon Common, London SW 19, will provide all details on where to see rural craftsmen, museums and collections. Their excellent publication, CoSIRA Guide to Country Workshops in Britain lists over a thousand places which welcome visitors.

The Museum of English Rural Life at the University of Reading has a comprehensive display of crafts information and the Welsh Folk Museum at St Faggans Castle, Cardiff, is outstanding. The West Yorkshire Folk Museum, Shibden Hall, Halifax features a variety of crafts as does the Weald and Downland Open-air Museum, Singleton, Chichester, Sussex. Good Open-air museums also exist at Avoncroft, near Bromsgrove, Stowmarket in Suffolk, and Hutton-le-Hole, Yorkshire. Other notable collections are at Bicton Gardens, Devon and Hartlebury, near Kidderminster, in addition to which, most local museums have collections of crafts that were practised in their areas. There are many private collections of items concerned with the old craftsman, some of which are open to the public. In general though, I would advise going to country areas, stopping at the village pub or shop and simply finding out if there are any craftsmen still working in the locality, for it is only by seeing these people at work, hearing them explain the operation in their own words, that a clear picture emerges. I hope you find it as interesting as I have.

Index